not associated or affiliated with them in any way. Nor does the referred product, website, and company names sponsor, endorse, or approve this product.

COMPENSATION DISCLOSURE: Unless otherwise expressly stated, you should assume that the links contained in this book may be affiliate links and either the author/publisher/reseller will earn commission if you click on them and buy the product/service mentioned in this book. However, the author/publisher/reseller disclaim any liability that may result from your involvement with any such websites/products. You should perform due diligence before buying mentioned products or services.

This constitutes the entire license agreement. Any disputes or terms not discussed in this agreement are at the sole discretion of the publisher.

Introduction

One of the best things you can do to expand your business is to start a YouTube channel. It's the most popular video-streaming service boasting massive online traffic, and constant streaming services – and ultimately, it will maximize exposure and help your brand become recognized within your market.

YouTube has afforded myself an amazing opportunity to get my message out to more people, very quickly and effectively. It is one of the best marketing channels to push your content and/or product.

And it's the **second most-visited website on the Internet** with nearly 2 billion monthly users logging on to watch videos every day.

So clearly it's a great way to connect with your

audience and keep them engaged for longer periods of time, while encouraging repeat traffic and views.

For any business owner, it only makes sense to use this as a marketing tool to better position yourself in your market. If you don't, it's a pretty safe bet that you're leaving a great deal of exposure on the table. And if your competitors are working hard to create attention-grabbing content for their channels, then you'll find it even more of a struggle to persuade your market that you have more to offer.

Unfortunately, a lot of people struggle to set up their channels. They find the learning curve too steep, or they believe that the investment of time and money may not pay off and chalk it up to a risk not worth taking.

After all, building a channel does take time and effort. Not only do you need to research the kind of content that is likely to attract attention (and become "sticky), but you have to be consistent with creating fresh content and engaging with your audience.

Still, regardless of the fact that creating a stand out channel takes time, there is no mistaking just how powerful YouTube is as a marketing vehicle that can carry your message wider and deeper into your market than nearly any other social platform.

And if you're unsure about how to get started while minimizing the learning curve, then keep reading because this special report was designed just for you.

I'll walk you through the ins and outs of YouTube, the best times to upload videos, and even how to make sure you attract the most people to your channel.

There's a lot of information out there about how to get started with YouTube. It can be overwhelming, so I've done my best to condense some of the most important tips in a way that is easy to follow. That way you're equipped with the basic knowledge you need to create your own channel.

Then, you can continue learning about YouTube including advanced optimization and marketing strategies later, once you have hands-on experience building and managing your channel.

So, don't complicate the process and try to learn everything about marketing on YouTube at once.

Take your time and focus on the basics: building a channel based on market research so you know what kind of content is in demand. Then optimize and expand!

One important thing to keep in mind is that you'll need to treat your YouTube channel as an extension of your business, not as just another social media platform. Building a successful YouTube channel takes time and effort and above all else, consistency.

That's the only way you will be able to gain traction and take your channel to the level you want.

Are you ready to get started?

Let's begin!

Tip #1: Have a Clear Vision

One of the most important things you can do is to develop a clear vision and strategy for your channel long before you begin to create content.

You want your channel to appeal to a specific audience so that you're able to connect with your core market. You also want visitors to immediately understand what you have to offer and that your content appeals to their interests.

This is how you connect to your audience and quickly build a large following on YouTube. All of your content should be unique, highly targeted and relevant to the products and services you are offering.

This requires careful content planning. You've probably already decided on a general focus for your channel, but now it's time to dig a bit deeper and create a content plan that will help you stay on track while ensuring that your channel is focused on what your audience wants most.

The key to YouTube is being consistent in terms of the type of content you offer as well as uploading fresh, unique content on a regular basis. Creating a content plan helps you stick to that overall theme.

Tip #2: Create an Attention-Grabbing Description

Believe it or not, a lot of new YouTube channels often overlook the importance of a highly-optimized channel description.

This is the first thing that people often see and if someone is interested in subscribing to your channel, chances are, they're going to read your description to determine whether it's something worthy and relevant to what they are most interested in.

Therefore, creating a well-written, keyword optimized, informative channel description is imperative to the success of your YouTube channel. Potential subscribers will find this when they click on your "About" tab.

To begin creating your channel description, click the "Edit" button within your admin panel, and then fill out the "About" section. Use this opportunity to inform viewers about the type of content you offer. Include links to your website or landing page so you can convert that viewer into a mailing list subscriber as well as a channel one!

You'll want to include 1-2 primary keywords that clearly describe your content. Use them in both video descriptions and titles as well. It's been proven that YouTube places a great deal of importance on the content found on your About Page so you want to make sure you use this space wisely.

There are also places to link your other social media accounts for your business, such as Facebook, Instagram, and Twitter, so be sure to fill them out if they're applicable, as this is a great opportunity to gain followers on other social media platforms and create a flow of traffic to all your accounts.

If your channel will offer a weekly podcast, then be sure to include the link and content schedule in your "About" section along with a relevant hashtag for your podcast.

Use YouTube as a spring board to generate and control traffic flow to other avenues of your business. It's one of the easiest ways to position yourself in your market and build a loyal following.

Once you've created a handful of videos, you can also create a playlist that is linked and featured in your About section. This is great for the times where you are offering longer videos on external channels like Twitch.

You can upload a portion of the full video to YouTube and then funnel traffic to the full video on another site where perhaps you're able to monetize.

At the end of your about section, you'll want to include something to encourage viewers to interact with your videos – essentially, a call to action.

Videos with a high level of interaction tell potential subscribers that what you produce is good, binge-worthy content, and keeps people engaged instead of clicking on something else and moving on.

For example, say something along the lines of something you've likely seen a hundred times on other channels:

"If you enjoy watching our videos, then please comment, hit that like button, subscribe to our channel, and share!"

This is an effective way to make sure people like, watch and subscribe. Most people are nervous at first to ask, but you must ask. This is the best way. People will subconsciously subscribe and like a video without thinking twice if you tell them.

Tip #3: Optimize Everything

A critical part of having a successful YouTube channel is Search Engine Optimization (SEO). This is what puts your videos on the Google map, as well as other search engines and social media sites, where people searching the internet can find it.

Be sure to familiarize yourself with Google's Keyword Planner, which is found at: https://ads.google.com/home/tools/keyword-planner/

You can also use Google Trends here too: https://trends.google.com

Both of these sites are effective when figuring out the best titles for your videos and the best keywords to use on each video.

Begin by evaluating different keyword phrases to determine what people are actively searching for, but also what keywords are relevant to your market. Chances are you'll uncover a ton of high-traffic keywords you haven't even considered.

Use a few of these keywords throughout your entire channel, not just your description but also in your video descriptions and title tags.

Keywords are what people search for when looking for specific content, so the more often you are incorporating them into your channel's content, the higher number of social signals will be out there, driving traffic in.

Also keep in mind that repetition matters. If you stay consistent with your channel, you will see consistent results.

Don't keyword stuff (this is where you use keywords in ways that don't make sense to your audience or detract from a viewer's experience).

At the same time, don't be afraid to use the same keywords multiple times throughout different areas of your channel. 2-3 times is best practice. Any more than that can get your account flagged.

Tip #4: Commit to A Schedule

It's worth saying again: *consistency is key when it comes to creating a successful YouTube channel.*

This means that you'll want to create a content plan that involves uploading content on specific days of the week. I remember starting and uploading only when I felt like it. It did nothing. My channel did not grow, nothing happened. But then I started getting consistent. More and more people were commenting. More and more people were subscribing. It was the best feeling.

You'll also want to make sure it's a schedule you can stick to since your subscribers will begin to rely on content updates on those days.

For example: Choose to publish weekly or daily

intervals such as every Monday, or Monday, Wednesday and Friday.

When you are first getting the hang of YouTube, but want to put out as much quality content as possible, see what you can realistically commit to per week.

In the beginning, you'll want to quickly build a backlist of videos so that you're able to keep people engaged when they arrive on your channel.

For example, you may find that you can only publish once or twice a week.

Decide on the days that you can guarantee videos be uploaded, and then put that information in the About section of your YouTube channel. This will tell viewers when they can expect to see a new video from you.

If you have other social media accounts, it's a good idea to mention which weekdays viewers can head on over to your channel for new content.

There's also a strategy for what time of day to publish.

Research shows that most people watch videos in the evenings and on weekends. But that doesn't mean you should upload during those times or expect a lot of views.

You want to give Google enough time to actually index your videos, meaning they're more likely to show up in search results.

During the week, it's suggested that you upload between 2-7pm EST. Wednesday is known to be the best day to upload (don't ask me why).

This gives your video enough time to be indexed for the viewers who will be sitting down that evening to watch. On Saturday and Sunday, they suggest publishing between 9am-11am.

Tip #5: Vary Video Type, But Focus

Now it's time to determine the types of marketing videos that you want to create.

Keep in mind that you need a good balance of variety while still sticking to your overall theme. The last thing you want is to confuse subscribers who are used to finding one type of content on your channel with a whole other thing.

On the other hand, viewers enjoy a variety of video content, and there are different ways that you can produce quality content while still sticking to your theme and brand.

Depending on the type of channel you are planning to create, you'll be able to offer a variety of content.

For example, if your channel is being created to promote other products and services, consider creating case-study style videos that highlight the progress your existing customers have made.

You could easily mix those types of videos in with product highlight videos, product review videos and even unboxing videos if you sell physical products.

Another type of marketing video is a How-To, DIY, or Tutorial. This is a good idea for when you want to demonstrate to potential buyers exactly how to use one of your products.

If your company sells home furnishings, then a How-To video about how to arrange furniture is a great way to showcase your products and educate the consumer.

Another good marketing video idea is a Listicle. These are especially popular for when you have several points to cover.

For example, a baking company may want to create a listicle video that discusses their most popular bakery treats. They can go over each dessert they offer, and then next to it on the screen list, promote an event that it would be ideal for.

And for practically any type of company, a funny, behind-the-scenes, candid video is sure to engage viewers. Show some bloopers from videos you've created, how you come up with video ideas, and anything else that helps viewers see the human side of your business.

For example, a telecommunications company could shoot a video of their customer service department.

In between shots of representatives helping customers, some employees could be shown having fun. This is a great way to humanize your company and connect with your audience on a deeper level.

Tip #6: Understand Your Audience

In order to both maintain and grow your YouTube audience, it's imperative that you know all about your current viewers and overall market. You'll want to be able to create a **visual snapshot** of who your average customer is.

In truth, market research should begin long before you create your YouTube channel. The more you know about your target audience, the easier it will be to create engaging content that they'll be interested in.

When you're inside of your YouTube admin panel, head on over to the Analytics tab.

This is going to give you a tremendous amount of information concerning your content and the type of visitors you're getting as well as what they are spending time watching.

On this page, you will find:

View Counts- the number of times your videos are being watched;

Average Watch Time- how long people stay on each video;

Revenue Generated- money earned;

Interaction Rate- physical engagements with content.

Pay close attention to the demographics section as well, because this will tell you:

Location in the world that your videos are being played;

Age range of viewers;

Gender of viewers.

Use this information to help your channel grow by knowing who your market is and what they want from you.

For example, make sure that your videos are being watched where you primarily offer your services. If they aren't, then think about marketing video ideas that can help localize your content.

Depending upon the type of company you own, you'll want to make sure that the age group and genders it's meant for are also watching your content. If they aren't, then come up with some new marketing videos that target these specific groups.

Once you review all the available information, head on over to the comments section on each of your videos. Take some time to see what kind of comments your videos are getting, and reply to ones that will help educate your viewers and promote your channel.

And finally, the YouTube Community profile is a fairly new feature that's starting to become quite popular. It's basically a timeline for you to post and interact with your followers.

This is where you'll want to post pictures that are relative to your channel, polls that will help you understand your audience, and even some fun, engaging content.

Consistently studying your analytics is the best way to improve your channel quickly. If you know why and why not the reason a video is performing a certain way, you can tweak or change the next video.

Tip #7: Know Your Competition

Just like your business does in the real world, you should keep tabs on your competitors on YouTube. Not only can they help you better understand your audience, but you'll be able to find inspiration from the successful channels in your market.

Set some time aside each week to review a handful of top channels in your niche.

Pay attention to the videos that receive both the highest and lowest number of views. Watch them yourself and pay attention to the different elements included in the video as well as the general style.

- What are they doing that you aren't doing?
- How could they improve their videos?

- What did you personally enjoy?
- What did you personally dislike?

All this information will help you improve, as well as shape your content so that it will likely appeal to most your market.

Never copy someone else's content! Just use them as a case-study and as inspiration for your own content ideas.

And just as you do for your own videos, take time to scroll through their comments. See how viewers respond, both negatively and positively. Perhaps someone has even referenced your brand in these comments.

Take notes of their descriptions and titles, what keywords they use, and see if you can apply it to your own content. Connect with people in your niche. This is the quickest and best way to grow. No one is your competition when you know everyone in your niche. Develop good rapport with other channels and this can turn to profitable JV(joint ventures).

Tip #8: Learn from Your Favorite Channels

Another great way to learn about YouTube success is during your downtime, when you can relax and watch videos from some of your own favorite channels.

These aren't the channels of your competitors, we've already covered that. This is when you spend some time in off-market channels, meaning that you're studying how other channels outside of your niche are increasing engagement and driving in traffic.

Just because they aren't in your niche doesn't mean you can't learn something valuable from them.

In fact, it's quite the opposite. By studying other markets outside of your own, you'll likely be able to generate fresh, new ideas that your own niche hasn't yet seen.

You'll also begin to see a pattern based on the content style that people are more excited about as well as how the channel owner chose to produce that content and in what formats (candid, professional, how-to style, etc.)

So, the next time you watch a video from one of your favorite creators, ask yourself a few questions.

For example, what about this video is holding my attention? Remember that YouTube is all about clicking, clicking, and clicking.

If this is a channel that you've been subscribed to for a while, then what keeps you coming back? Is it the graphics, variety of videos, tone, style, or something else?

Look at the techniques they use to keep viewers returning day after day. Perhaps they include giveaways, fun facts, candid interviews, or live streaming sessions. Are any of these something that you can do on your own channel?

Get those creative juices flowing! Never create the same type of video twice if the first video did not perform. Also, don't get stuck on a video that does not perform. Sometimes it happens.

A thumbnail is the picture that you click on when you want to watch a video. It should clearly indicate what the video is about, even though your title and description should also reflect the topic of your video content. If you don't have a good thumbnail, or a thumbnail that is not authentic, then people will not click. You can see this in the impressions tab of your statistics for your channel.

One option you have is to choose a freeze-frame from the video that you've uploaded.

For example, if you're a baking company who also sells products to bake with, you might upload a video on how to bake brownies.

You can choose a freeze-frame from the video that shows you mixing the batter. YouTube automatically generates a few that they think are good ones, so this is fairly easy to do.

But the more popular option is to create a custom thumbnail and then upload it, as you can incorporate something that may not be in the video but will entice people to click on it.

This is as simple as taking a picture and using a picture editing tool, such as Photoshop, or free online editing tools like https://www.Canva.com, to make it more appealing and relatable to your video.

For example, a beauty brand may want to use a custom thumbnail that features the products being used in the video hovering around a person's face, using arrows to point where the products will be applied.

Or you can do a split-screen custom thumbnail. In the case of this beauty channel, they can use a before and after picture of what someone looks like with and without makeup, side by side.

Be sure to play around with different looks before deciding on your custom thumbnail, and make sure that it's interesting enough that even you would click on it. Testing different thumbnails can help you see which work, and which don't.

In addition to everything else, it's important to have both a profile picture and channel art on your YouTube channel that accurately represents your company and reflects your brand message.

And if your business has accounts on other social media platforms, then your business profile picture and channel art should be the same. Again, it's all about consistency!

Your profile picture and channel art are the very first things that people will see when they visit your channel. Whether you're a one-person business or a company with thousands of employees, it's important for brand recognition.

Your profile picture, for individual businesses, can be a professional picture of yourself, a mascot or another image that works with your brand. Simply click on the picture within your channel to upload and edit.

Your channel art, however, should be a bit more brand-specific.

You may want to consider hiring a graphic designer to create your logo or channel art. Or if your brand already has a logo, simply upload and crop accordingly.

Keep in mind what colors your brand uses the most, as well. Then try and incorporate these colors into your videos as often as possible. For example, if your company's logo is cobalt blue, then have videos with a splash of this color in it, or use other graphic elements like a video bar or subscribe button that includes that theme.

When visitors go to your channel and see your cobalt blue logo above thumbnails with matching thumbnails, they'll start to associate that color with your brand. If you've spent some time analyzing successful YouTube channels, chances are you've noticed how common it is for all thumbnails to look similar in terms of color or font styles used.

This helps to build a strong visual identity that helps your account look polished and professional, but also helps to establish a theme. Building a brand around your channel and staying consistent with that brand is the most important part of long-term success on Youtube.

Tip #10: Sign Up for G-Suite account

YouTube is owned by Google, and so your channel will automatically be linked with a Gmail account. Even if you already have one, it's a good idea to create a **separate account** just for your business.

Look into investing in G Suite in order to have professional email addresses created. So rather than @gmail.com, your email will say @yourcompanyname.com. It's a small step, but it helps promote a professional atmosphere for your channel and company overall.

Keep in mind that the name you choose for your new Google account should be the name of your brand.

The next few steps are self-explanatory, until you get to where it asks you to setup a YouTube brand account. This will allow you to designate editing positions.

Just head on over to YouTube, click on your profile icon in the upper right hand corner, and scroll down to "My Channel." Click on it, but avoid clicking on "Create Channel."

This is where you want to click "Use a business or other name," down at the bottom. At this point you should be prompted to enter your *Brand Account* name.

Your Brand Account name can always be changed in the future, should you decide to do so. At this point you have successfully set up the foundation for your company channel.

Final Words

I hope that this special report has given you the direction you need to start building your very own YouTube channel.

Start off by researching your market, spend some time watching videos from successful channels in your niche and then develop a content plan and schedule that you can stick to.

Once you get the hang of it, make it a point to regularly upload quality content based on the schedule you've chosen. Try not to change this!

Your subscribers will rely on it and begin to log on expecting new content on those days.

Give viewers a reason to subscribe to your channel and keep coming back, and don't forget to directly ask for it! A call to action within each video as well as channel description is important, but you should always encourage them to also turn on notifications so that they're alerted whenever you upload fresh content.

Notifications cuts through the noise and helps your channel gain traction with repeat visits from subscribers.

And remember to upload in different video formats to keep it interesting, and keep the viewer's attention.

We wish you the best of luck in creating a quality channel and hope you've enjoyed this special report!

I wish you much success on your journey through YouTube. I will not deny the saturation levels for YouTube is increasing, but I will not say that it is too saturated.

Have a strong message to share, be unique in your message delivery, and shock your audience, each and every video.

The channels that end up getting millions of subscribers are the channels with the best editors.

Resources to check out:

How to Start Your YouTube Channel

>>https://adespresso.com/blog/youtube-marketing/

YouTube Marketing Tips:

>>https://www.wordstream.com/blog/ws/2018/09/20/youtube-marketing

Growing Your YouTube Channel:

>>https://blog.hootsuite.com/youtube-marketing/

HubSpot Tips:

>>https://www.hubspot.com/youtube-marketing

YouTube Tips:

>>https://neilpatel.com/blog/youtube-marketing-guide/

How To Market To Your Audience:

>>https://backlinko.com/youtube-marketing

Cheat Sheet For YouTube Channel Marketing:

>>https://buzzsumo.com/blog/youtube-for-popular-content-the-complete-beginners-guide-to-youtube-marketing/https://www.dummies.com/social-media/youtube/youtube-marketing-for-dummies-cheat-sheet/

Create Thumbnails Easily:

https://www.Canva.com

www.ingramcontent.com/pod-product-compliance
Lightning Source LLC
Chambersburg PA
CBHW031501210526
45463CB00003B/1019